International Mammalian Synthetic Biology Workshop (mSBW 2020)

Online
7-8 December 2020

ISBN: 978-1-7138-2772-6

Printed from e-media with permission by:

Curran Associates, Inc.
57 Morehouse Lane
Red Hook, NY 12571

proceedings.com

Some format issues inherent in the e-media version may also appear in this print version.

Printed with permission by Curran Associates, Inc. (2021)

For permission requests, please contact AIChE
at the address below.

AIChE
120 Wall Street, FL 23
New York, NY 10005-4020

Phone: (800) 242-4363
Fax: (203) 775-5177

www.aiche.org

Additional copies of this publication are available from:

Curran Associates, Inc.
57 Morehouse Lane
Red Hook, NY 12571 USA
Phone: 845-758-0400
Fax: 845-758-2633
Email: curran@proceedings.com
Web: www.proceedings.com

TABLE OF CONTENTS

GENERAL SUBMISSIONS

Human Cells-Produced Metallic Nanomaterials: A Look Beyond Biomedical Applications.

Conference: Mammalian Synthetic Biology Workshop

Year: 2020

Proceeding: Virtual International Mammalian Synthetic Biology Workshop (mSBW) 2020

Group: General Submissions

Session: General Submission (Oral)

Authors:

Truong, L., Northeastern University
Medina, D., Northeastern University
Vernet-Crua, A., Northeastern University
Webster, T. J., Northeastern University

In this study, we will clarify ph treated (Ph 13 > Ph5) E.Coli lysate with a single use centrifuge UniFuge®. Day one is to determine feasibility of two application scenarios, The fermenter generates 16 liters of E. Coli . The pcv is approx. 10% after acid treatment. Day one, we will determine shear sensitivity thru pump and UniFuge. We will apply low shear parameters- low g. Day two, we will process 34 liters in 5 trials. Lysate will be PH 5 prior to pump.

Objective- Collect centrate, soluble Mrna for Covid 19 vaccine. This feed stream is reputed to be a shear sensitive process. We will apply low shear parameters to determine the best sequence by NTU, and gel electrophoresis. Longer term analytical results will be performed to check for contamination and potency.

In this study, we will clarify ph treated (Ph 13 > Ph5) E.Coli lysate with a single use centrifuge UniFuge®. Day one is to determine feasibility of two application scenarios, The fermenter generates 16 liters of E. Coli . The pcv is approx. 10% after acid treatment. Day one, we will determine shear sensitivity thru pump and UniFuge. We will apply low shear parameters- low g. Day two, we will process 34 liters in 5 trials. Lysate will be PH 5 prior to pump.

Reconstitution of Prospermatogonial Specification in Vitro from Human Induced Pluripotent Stem Cells

Conference: Mammalian Synthetic Biology Workshop

Year: 2020

Proceeding: Virtual International Mammalian Synthetic Biology Workshop (mSBW) 2020

Group: General Submissions

Session: General Submission (Oral)

Authors:

Sasaki, K., University of Pennsylvania

The ability to insert a transgenic payload into a specific locus in a host cellâ€™s chromosome facilitates the creation of cell lines that reliably express therapeutic transgenes or synthetic gene circuits. Such targeting can be achieved with bacteriophage-derived serine integrases â€“ enzymes that catalyze recombination between short DNA sequences, called *attP* and *attB* sites. Previous work has shown that the Bxb1 integrase works efficiently in mammalian cells. Here, we show that a point mutation in the Bxb1 *attP* and *attB* sites, termed Bxb1-GA, nearly doubles the efficiency of Bxb1-mediated integration, resulting in the highest integration efficiency reported to date in mammalian cells. Bxb1-GA sites and wild-type Bxb1 sites do not cross-react, and hence can be used in parallel for orthogonal targeting of distinct payloads to different sites. Our data present an improved tool for site-specific transgene integration in mammalian cells for research and biomanufacturing purposes.

Optogenetic Control of Apical Constriction Induces 3D Deformation in Mammalian Epithelia

Conference: Mammalian Synthetic Biology Workshop

Year: 2020

Proceeding: Virtual International Mammalian Synthetic Biology Workshop (mSBW) 2020

Group: General Submissions

Session: General Submission (Oral)

Authors:

Martínez Ara, G., EMBL Barcelona
Taberner Carretero, N., RIKEN BDR
Takayama, M., RIKEN BDR
Sandaltzopoulou, E., EMBL Barcelona
Ebisuya, M., EMBL Barcelona

During development, cellular forces come together in a synchronized manner to robustly generate functional tissue shape. Apical constriction is one of these force generating processes, and has been demonstrated to be necessary for the induction of epithelial curvature in fundamental morphogenetic events, such as neural tube folding[1]. The emerging field of synthetic developmental biology proposes the use of bottom-up approaches to learn how such processes combine to create functional tissues[2]. However, to this date thereâ€™s no tool available to manipulate 3D shape of mammalian tissues. Here we report the development of â€œOptoShroom3â€□, a new optogenetic tool that achieves fast spatiotemporal control of apical constriction in mammalian epithelia. We demonstrate that OptoShroom3 induced apical constriction can provoke tissue curvature and folding on an MDCK cell sheet. Moreover, its application in eye-cup organoids lead to epithelial thickening and apical lumen reduction in optic vesicles. These results show that spatiotemporal control of apical constriction can be used to generate controlled 3D tissue shape changes in mammalian epithelia. OptoShroom3 also points out the potential of optogentics to gain control over morphogenetic processes and develop tissue architectures in vitro.

1. Haigo, S. L., Hildebrand, J. D., Harland, R. M., & Wallingford, J. B. (2003). Shroom induces apical constriction and is required for hingepoint formation during neural tube closure. Current biology, 13(24), 2125-2137.

2.Ebrahimkhani, M. R., & Ebisuya, M. (2019). Synthetic developmental biology: build and control multicellular systems. Current Opinion in Chemical Biology, 52, 9-15.

Rapid Directed Molecular Evolution of Fluorescent Proteins in Mammalian Cells

Conference: Mammalian Synthetic Biology Workshop

Year: 2020

Proceeding: Virtual International Mammalian Synthetic Biology Workshop (mSBW) 2020

Group: General Submissions

Session: General Submission (Oral)

Authors:

Babakhanova, S., Stanford
Jung, E., University of Illinois at Chicago
Namikawa, K., Technische Universität Braunschweig
Subach, O., National Research Center “Kurchatov Institute”
Korzhenevskiy, D., National Research Center “Kurchatov Institute”
Rakitina, T., National Research Center “Kurchatov Institute”
Xiao, X., Westlake University
Cosio, D., Media Lab, Massachusetts Institute of Technology
Shi, J., Westlake Laboratory of Life Sciences and Biomedicine, Westlake University
Wu, S., Westlake Laboratory of Life Sciences and Biomedicine, Westlake University
Park, D., Media Lab, Massachusetts Institute of Technology
Eisenhard, L., Division of Cellular and Molecular Neurobiology, Zoological Institute, Technische Universität Braunschweig
Tang, H., Westlake Laboratory of Life Sciences and Biomedicine, Westlake University
Yi, S., Westlake Laboratory of Life Sciences and Biomedicine, Westlake University
Köster, R., Division of Cellular and Molecular Neurobiology, Zoological Institute, Technische Universität Braunschweig
Subach, F., National Research Center “Kurchatov Institute”
Piatkevich, K., Westlake Laboratory of Life Sciences and Biomedicine
Boyden, E. S., Massachusetts Institute of Technology

In vivo imaging of model organisms is heavily reliant on fluorescent proteins with high brightness. Here we describe a practical method for rapid optimization of fluorescent proteins via directed molecular evolution in mammalian cells. We employed this approach to enhance intracellular brightness of a set of the spectrally diverse fluorescent proteins. The developed proteins demonstrated high performance for fluorescent labeling of neurons in culture and *in vivo* in model organisms such as *C.elegans*, *Drosophila* fruit flies, and zebrafish. The described method has great potential to be adopted by protein engineers due to its simplicity and convenience. We also believe that the new enhanced fluorescent proteins will find wide application for *in vivo* imaging of small model organisms.

Single Use Centrifuge < Unifuge for Cell Collection /Wash/Buffer Exchange

Conference: Mammalian Synthetic Biology Workshop

Year: 2020

Proceeding: Virtual International Mammalian Synthetic Biology Workshop (mSBW) 2020

Group: General Submissions

Session: General Submission (Oral)

Authors:

Richardson, D. R., PSangelus

In this study, we will clarify ph treated (Ph 13 > Ph5) E.Coli lysate with a single use centrifuge UniFuge®. Day one is to determine feasibility of two application scenarios, The fermenter generates 16 liters of E. Coli . The pcv is approx. 10% after acid treatment. Day one, we will determine shear sensitivity thru pump and UniFuge. We will apply low shear parameters- low g. Day two, we will process 34 liters in 5 trials. Lysate will be PH 5 prior to pump.

Objective- Collect centrate, soluble Mrna for Covid 19 vaccine. This feed stream is reputed to be a shear sensitive process. We will apply low shear parameters to determine the best sequence by NTU, and gel electrophoresis. Longer term analytical results will be performed to check for contamination and potency.

In this study, we will clarify ph treated (Ph 13 > Ph5) E.Coli lysate with a single use centrifuge UniFuge®. Day one is to determine feasibility of two application scenarios, The fermenter generates 16 liters of E. Coli . The pcv is approx. 10% after acid treatment. Day one, we will determine shear sensitivity thru pump and UniFuge. We will apply low shear parameters- low g. Day two, we will process 34 liters in 5 trials. Lysate will be PH 5 prior to pump.

From Sequence to Yield: Deep Learning for Protein Production Systems

Conference: Mammalian Synthetic Biology Workshop

Year: 2020

Proceeding: Virtual International Mammalian Synthetic Biology Workshop (mSBW) 2020

Group: General Submissions

Session: General Submission (Oral)

Authors:

Oyarzún, D., Imperial College London
Nikolados, E. M., University of Edinburgh

A key area in biotechnology is the production of recombinant proteins for the energy, food and pharmaceutical sectors. Protein expression systems are typically engineered to produce large amounts of protein, but this requires many iterations between strain design, construction and characterisation. Here we propose an advanced machine learning pipeline to forecast production and improve performance. Our algorithms can predict protein production from DNA sequence with >80% accuracy. We trained machine learning models of increasing complexity, from simple linear regressors to deep neural networks, on large phenotypic screens of GFP readouts and fitness. Our results reveal advantages and caveats of various algorithms and highlight trade-offs between prediction accuracy, the size of the training dataset, and the type of algorithms employed. We demonstrate that the use of Transfer Learning can decrease the data requirements by up to five fold when training on multiple phenotypes. Our results lay the groundwork for machine learning-enabled engineering of strains with improved yields, providing a unique, data-driven, approach to accelerate strain design and optimisation.

High-Efficiency System for Site-Specific Transgene Integration in Mammalian Cells

Conference: Mammalian Synthetic Biology Workshop

Year: 2020

Proceeding: Virtual International Mammalian Synthetic Biology Workshop (mSBW) 2020

Group: General Submissions

Session: General Submission (Oral)

Authors:

Jusiak, B., MIT
Jagtap, K., MIT
Gaidukov, L., MIT
Zhang, L., Pfizer
Weiss, R., M.I.T.
Lu, T. K., MIT

The ability to insert a transgenic payload into a specific locus in a host cellâ€™s chromosome facilitates the creation of cell lines that reliably express therapeutic transgenes or synthetic gene circuits. Such targeting can be achieved with bacteriophage-derived serine integrases â€“ enzymes that catalyze recombination between short DNA sequences, called *attP* and *attB* sites. Previous work has shown that the Bxb1 integrase works efficiently in mammalian cells. Here, we show that a point mutation in the Bxb1 *attP* and *attB* sites, termed Bxb1-GA, nearly doubles the efficiency of Bxb1-mediated integration, resulting in the highest integration efficiency reported to date in mammalian cells. Bxb1-GA sites and wild-type Bxb1 sites do not cross-react, and hence can be used in parallel for orthogonal targeting of distinct payloads to different sites. Our data present an improved tool for site-specific transgene integration in mammalian cells for research and biomanufacturing purposes.

Temperature-Responsive Optogenetic Probes of Ras and PI3K Signaling

Conference: Mammalian Synthetic Biology Workshop

Year: 2020

Proceeding: Virtual International Mammalian Synthetic Biology Workshop (mSBW) 2020

Group: General Submissions

Session: General Submission (Oral)

Authors:

Bugaj, L., University of California, San Francisco
Benman, W., University of Pennsylvania - Philadelphia, PA
Berlew, E., University of Pennsylvania
Chow, B., University of Pennsylvania

Optogenetic signaling probes have served as powerful tools to dissect the role of spatiotemporal signal dynamics in cells and organisms. However, such tools can be challenging to implement because they often consist of multiple components that require stoichiometric tuning and occupy valuable fluorescence channels. We thus sought to create single-component variants of optogenetic probes that work through membrane recruitment, a common mode of optogenetic control. Specifically, we engineered control of Ras/Erk and PI3K/Akt signaling by fusing pathway activators to BcLOV4, a recently characterized photoreceptor that translocates to membrane lipids upon photoactivation. Our probes allowed robust activation of both pathways and compared favorably to existing 2-component probes in terms of basal signal perturbation and dynamic range of induction. However, we were surprised to discover that BcLOV4 responds not only to light, but also to temperature, such that membrane localization and pathway stimulation spontaneously decay â€"despite constant illumination â€" at a rate proportional to both the temperature and the intensity of light. We systematically characterized this unique dependency on both light and temperature, and we developed a computational model that fully predicts BcLOV4 translocation and Ras/Erk activation dynamics. Our model reveals that BcLOV4 permits stable activation at lower temperatures (< 30C), but activation decays at higher temperatures. Our model also provides a predictive model for the use of BcLOV-based tools. In addition to use in mammalian cells, we show that BcLOV4-based tools function robustly and stably in *Drosophila* and zebrafish, which are not amenable to stoichiometric expression tuning, and which are cultured at permissive temperatures. Ongoing protein engineering efforts will permit tuning of the temperature-dependence of BcLOV4 for its application across an even wider range of experimental systems and conditions.

Development of a Toggle Switch Based on Post-Transcriptional Regulation That Resists Epigenetic Silencing

Conference: Mammalian Synthetic Biology Workshop

Year: 2020

Proceeding: Virtual International Mammalian Synthetic Biology Workshop (mSBW) 2020

Group: General Submissions

Session: General Submission (Oral)

Authors:

Wauford, N., MASSACHUSETTS INSTITUTE OF TECHNOLOGY
DiAndreth, B., MIT
Weiss, R., M.I.T.

Synthetic biological circuits may enable sophisticated cell therapies that respond to patient biomarkers and inputs from clinicians. However, transcriptionally regulated genetic circuits are prone to epigenetic silencing, in which chromatin modifications often turn off transgene expression after a few weeks. This poses a significant challenge to the regulation of cell therapies that may persist *in vivo* for years. Constitutive transcription has been shown to reduce silencing, but tools to regulate gene expression post-transcriptionally are limited. To address this gap, we present the PERSIST platform for post-transcriptional regulation, in which RNA cleavage events activate or repress downstream gene expression. We take advantage of CRISPR proteins in the Cas6 and Cas13 families that cleave sequence-specific RNA hairpins, demonstrating a set of nine proteins that enact activation and repression, often with fold changes of two orders of magnitude. PERSIST genetic circuits show 28% less silencing than the transcriptionally regulated TetON system after 1 month and no more silencing than a constitutive control. We then used the PERSIST platform to implement a genetic toggle switch. Toggle switches could improve cell therapies by imparting them with memory, such that a transient input, such as a drug input from a clinician, can permanently change cell behavior.

We designed a system in which FDA-approved small molecule drugs could switch a cell between two behavioral states, and implemented this system by combining PERSIST logic processing with engineered small molecule-responsive fusion proteins. Preliminary results suggest that our circuit is able to switch between two stable states in response to clinically relevant small molecule drugs. These results will provide a platform to regulate the expression of transgenes in cell therapies over time, enabling optimized dosing schedules of therapeutic proteins in applications including encapsulated cell therapies, engineered immune cells, and gene therapies.

Modular RNA-Responsive Elements for Versatile Eukaryotic Translational Control

Conference: Mammalian Synthetic Biology Workshop

Year: 2020

Proceeding: Virtual International Mammalian Synthetic Biology Workshop (mSBW) 2020

Group: General Submissions

Session: General Submission (Oral)

Authors:

Zhao, E., Princeton University
Collins, J. J., Massachusetts Institute of Technology

RNA-responsive elements enable reprogramming of biological systems. Although much progress has been made in this field, eukaryotic systems still lack simple RNA responsive elements that can be placed directly in front of a target translationally controlled gene. Such a system would allow RNA therapies to be more specific based on endogenous RNA content and open up further possibilities for cellular therapies. Here, we introduce eukaryotic toehold switches (eToeholds) as a modular, ~300-500 base pair long, riboregulator that functions in eukaryotic systems. eToeholds were designed through meticulous optimization of RNA base pair mediated folding and expression methodology. eToeholds function within a range of eukaryotic organisms and may be used to sense and respond to the presence of a selected trigger RNA with up 16-fold induction of any transgene. We show a powerful application of eToeholds in the sensing of viral infection, where a Zika-targeted eToehold increases the regulated expression of luciferase 9-fold higher than controls after Zika infection. We finally demonstrate that eToeholds can be used to sense both endogenous transcript changes and cell-type specific transcript differences. The robust and specific activation and the modular design of eToeholds demonstrate their broad applicability for numerous applications in biological engineering.

Cell-State Specific Digital to Analog Signal Conversion in Human Pluripotent Stem Cells

Conference: Mammalian Synthetic Biology Workshop

Year: 2020

Proceeding: Virtual International Mammalian Synthetic Biology Workshop (mSBW) 2020

Group: General Submissions

Session: General Submission (Oral)

Authors:

Prochazka, L., University of Toronto
Zandstra, P. W., University of Toronto
Michaels, Y., University of British Columbia
Lau, C., University of British Columbia
Kaul, H., University of British Columbia
Siu, M., University of British Columbia
Yin, T., University of Toronto
Wu, D., University of Toronto
Jang, E., University of Toronto
Benenson, Y., Swiss Federal Institute of Technology (ETH Zürich)

Human pluripotent stem cells (hPSC), due to their ability to self-replicate and differentiate *in vitro*, are an ideal cell source for generation of cell-based therapeutics and as a model system for drug screening and basic research. Controlling hPSC through synthetic gene circuits has the potential to tackle key challenges in the field and can increase our understanding of basic cellular and molecular mechanisms. To achieve advanced cell state control, regulatory- proteins or RNAs often need to be i) restricted to a desired cell state- and ii) produced at defined physiological amounts. While previous synthetic circuits have achieved either digital cell-state specific actuation or analogue output tuning, none have combined these functionalities in hPSC or other human cells. Here, we have created a first generation of digital to analog converter in hPSC that operate entirely on endogenous miRNAs. Our digital to analog converter is based on a bow-tie framework composed of a digital input module, that detects multiple endogenous miRNAs, integrates them in an AND-like logic function and, in turn, controls an analog output module controlled by a library of miRNA17 target site variants. To implement this circuit in hPSC, we used a previously established automated design toolbox to search for a set of miRNAs and a circuit topology that maximizes output expression in hPSC compared to other cell types in the testing space using published miRNA Seq data. We then optimize logic processing of the identified miRNAs using a model-guided approach and finally validate cell-state specific analog processing in hPSC. We then apply the circuit to fine tune BMP4, a morphogen that is required to induce peri-gastrulation like pattern formation *in vitro* and demonstrate BMP4-dose dependent pattern formation in the absence of BMP4 in the culture media.

Using Synthetic Biology to Generate Hyper-Stable Vaccines

Conference: Mammalian Synthetic Biology Workshop

Year: 2020

Proceeding: Virtual International Mammalian Synthetic Biology Workshop (mSBW) 2020

Group: General Submissions

Session: General Submission (Oral)

Authors:

Miles, J., James Cook University

Polypeptide vaccines effectively activate human T cells but suffer from poor biological stability, which confines both transport logistics and in vivo therapeutic activity. Synthetic biology has the potential to address these limitations through the generation of highly stable antigenic â€œmimicsâ€□ using subunits that do not exist in the natural world. We developed a platform based on non-natural chemistry and used this platform to reverse engineer fully artificial T cell agonists that immunogenicity more than 5-fold of their natural blueprints. This non-natural chemistry is highly stable in human serum and gastric acid. *In vitro*, these synthetic agonists expanded antigen-specific responses against multiple epitopes across multiple viruses. *In vivo*, synthetic vaccinated mice were protected from lethal challenge. Moreover, the synthetic agonists were immunogenic after oral administration. We have since expanded this technology to other human viruses including SARS-CoV-2. These proof-of-concept studies highlights the power of synthetic biology to expand the horizons of vaccine design and therapeutic delivery.

Reduction of Gene Expression Noise Caused By Plasmid Copy Number Variation in Mammalian Cells.

Conference: Mammalian Synthetic Biology Workshop

Year: 2020

Proceeding: Virtual International Mammalian Synthetic Biology Workshop (mSBW) 2020

Group: General Submissions

Session: General Submission (Oral)

Authors:

Lee, J., Rice University
Yang, J., MIT
Land, M., Baylor College of Medicine
Igoshin, O., Rice University
St-Pierre, F., Baylor College of Medicine

In vivo imaging of model organisms is heavily reliant on fluorescent proteins with high brightness. Here we describe a practical method for rapid optimization of fluorescent proteins via directed molecular evolution in mammalian cells. We employed this approach to enhance intracellular brightness of a set of the spectrally diverse fluorescent proteins. The developed proteins demonstrated high performance for fluorescent labeling of neurons in culture and *in vivo* in model organisms such as *C.elegans, Drosophila* fruit flies, and zebrafish. The described method has great potential to be adopted by protein engineers due to its simplicity and convenience. We also believe that the new enhanced fluorescent proteins will find wide application for *in vivo* imaging of small model organisms.

A Replicon Platform for Precise Reprogramming of the Tumor Microenvironment

- Conference: Mammalian Synthetic Biology Workshop
- Year: 2020
- Proceeding: Virtual International Mammalian Synthetic Biology Workshop (mSBW) 2020
- Group: Poster Session
- Session:

  General Submission

Authors:
Caliendo, F. - Presenter, Synthetic Biology Center, Department of Biological Engineering, Massachusetts Institute of Technology
Maiorino, L., Massachusetts Institute of Technology
Remba, T. K., Massachusetts Institute of Technology
Chadha, A. R., Massachusetts Institute of Technology
Irvine, D. J., MIT
Weiss, R., Massachusetts Institute of Technology
Among immunotherapy treatments under investigation, *in situ* vaccination with self-amplified RNA derived from alphaviruses (replicon) is a promising strategy, achieving superior efficacy in T-cell activation than conventional RNA and DNA approaches, by coupling persistent expression of therapeutic immunomodulatory genes to the stimulation of innate immunity at the tumor site. Regardless of the therapeutic payload expressed, the nature of the tumor microenvironment (TME) determines the responsiveness of cancer to the replicon vaccine. The TME, can be simply classified into hot or cold based on the levels of proinflammatory cytokines and T cell infiltration, with the former presenting the higher response rate and the latter resistance to cancer immunotherapy. Considering the pivotal role that the nature of the TME plays in determining the effectiveness of immunotherapy, strategies to convert cold tumors into hot ones and thus sensitize tumors to replicon vaccines are of great interest.

We are developing a next generation precision replicon platform to reprogram cold tumors into hot ones via delivery of replicon RNA molecules that selectively induce apoptosis in tumor cells and reprogram key immune infiltrates in the TME. To provide the required selectivity, we implemented a classifier circuit which consists of microRNA (miRNA) sensing modules that by sensing the miRNA profile of the recipient cell, determine the expression of the genetic cargo only in the desired cell type. We show that the presence of miRNA sensing modules aligns the expression of reporter genes to the miRNA profile of tumor and immune cell lines. Replicons equipped with miRNA sensor modules specifically express the encoded payload in primary murine $CD8^+$ T cells. Together, these results show the potential application of our replicon platform to precisely deliver synthetic circuits to specific cell types in the TME.

Design of Mammalian on-Riboswitches Based on Tandemly Fused Aptamer and Ribozyme

- Conference: Mammalian Synthetic Biology Workshop
- Year: 2020
- Proceeding: Virtual International Mammalian Synthetic Biology Workshop (mSBW) 2020
- Group: Poster Session
- Session:

  General Submission

Authors:
Mustafina, K. - Presenter, Okinawa Institute of Science and Technology
Fukunaga, K., Okinawa Institute of Science and Technology
Yokobayashi, Y., Okinawa Institute of Science and Technology Graduate University
Chemically regulated self-cleaving ribozymes, or aptazymes, are emerging as a promising class of genetic devices that allow dynamic control of gene expression in synthetic biology. However, further expansion of the limited repertoire of ribozymes and aptamers, and development of new strategies to couple the RNA elements to engineer functional aptazymes are highly desirable for synthetic biology applications. Here, we report aptazymes based on the recently identified self-cleaving pistol ribozyme class using a guanine aptamer as the molecular sensing element. Two aptazyme architectures were studied by constructing and assaying 17 728 mutants by deep sequencing. Although one of the architectures did not yield functional aptazymes, a novel aptazyme design in which the aptamer and the ribozyme were placed in tandem yielded a number of guanine-inhibited ribozymes. Detailed analysis of the extensive sequence-function data suggests a mechanism that involves a competition between two mutually exclusive RNA structures reminiscent of natural bacterial riboswitches.

A Resource Toolbox for Imaging Membrane Voltage of Mammalian Cells

Conference: Mammalian Synthetic Biology Workshop

Year: 2020

Proceeding: Virtual International Mammalian Synthetic Biology Workshop (mSBW) 2020

Group: Poster Session

Session: General Submission

Authors:

Lu, X., Rice Unviersity
Liu, Z., Rice University
Gou, Y., Baylor College of Medicine
Guan, S., Baylor College of Medicine
Lai, S., Baylor College of Medicine
St-Pierre, F., Baylor College of Medicine
Reimer, J., Baylor College of Medicine
Tolias, A., Baylor College of Medicine

Protein indicators had been widely used for monitoring biological signals in vivo, out of which Genetically Encoded Voltage Indicators (GEVIs) are a promising new technology for monitoring electrical activity in the brain noninvasively with high spatiotemporal resolution. GEVIs fundamentally work by changing their brightness in response to changes in membrane potential. As a relatively new technology, thus there is room for further optimization of GEVI to expand their versatility and applications. We present here a newly designed GEVI called Jellyfish-derived Electricity-reporting Designer Indicator (JEDI) with significantly improved brightness and responsivity as compared to other commonly used GEVIs. JEDIs can be readily expressed in several different host systems through plasmid or virus delivery, enabling a broad range of applications.

Untangling Resource Competition Towards Predictable Mammalian Cell Engineering

Conference: Mammalian Synthetic Biology Workshop

Year: 2020

Proceeding: Virtual International Mammalian Synthetic Biology Workshop (mSBW) 2020

Group: Poster Session

Session: General Submission

Authors:

Di Blasi, R., Imperial College London
Ceroni, F., Imperial College London
Siciliano, V., Italian Institute of Technology
Tedeschi, F., Istituto Italiano di Tecnologia (IIT)
Pisani, M., Istituto Italiano di Tecnologia (IIT)

Despite the many achievements in synthetic biology, building new engineered systems still poses numerous challenges, such as poor predictability and reliability. This is mainly because heterologous constructs rely on the very same pool of resources used by cells to sustain their physiology. As a consequence, synthetic constructs impose an extra load Ã¢â,¬â€œ also referred to as *burden*Ã¢â,¬â€œ to engineered cells that have to re-allocate their finite amount of cellular resources to match the transcriptional and translational needs of the engineered systems(Ceroni *et al.*, 2015; Borkowski *et al.*, 2016). This greatly reduces cellular gene expression and metabolic capacities, leading to unexpected behaviours and hampering a straightforward engineering(Tan, Marguet and You, 2009; Gyorgy *et al.*, 2015).

Here, we investigate the parameters to be considered when the design of low-burden synthetic construct for mammalian cell engineering is sought. As reported in previous findings(HuliÃƒÂik *et al.*, 2012; Frei *et al.*, 2020), we confirm that inducible expression of heterologous proteins heavily impacts the behaviour of a co-transfected constitutive reporter. This has already been pointed out as a consequence of resource competition arising between the two co-transfected modules(HuliÃik *et al.*, 2012; Frei *et al.*, 2020). We have also built a library of constructs with different intracellular localisation, transcriptional strength, and translational efficiency, showing how these three parameters are important when the tuning of such competition is sought.

We next aim to investigate the role played by burden when working with integrated circuits in mammalian cells, with the ultimate aim of individuating new design rules predictive of reduced burden. This will greatly benefit numerous bioproduction and biotherapeutic applications relying on mammalian cell engineering.

Serenity Stabilizers for Autistic Users

Conference: Mammalian Synthetic Biology Workshop

Year: 2020

Proceeding: Virtual International Mammalian Synthetic Biology Workshop (mSBW) 2020

Group: Poster Session

Session: General Submission

Authors:

Neff, B., Kyro Tech LLC

Serenity Stabilizers

This will be a hearing device for autistic users which will lower the decibel of sound in a room for users. This device will have a removable rechargeable battery. The main components will be a speaker, microphone, and an attenuator. This product will be the same size as a hearing aid and look exactly like one. This product will allow users to attend concerts comfortably as loud sounds hurt their ears. This product will come in three different colors cream, tan, and brown. There will be colorful covers available in silicone or felt fabric. The noise reduction rate will be atleast 36, and it will block out extraneous background noise and dampen loud noises yet you will still be able to hear whispers. The ear mold of the device will be covered in a soft foam similar to ear plugs. The price will be around $200 a set.

Monitoring Transient Tumour-Immune Cell Interactions Using Synnotch

Conference: Mammalian Synthetic Biology Workshop

Year: 2020

Proceeding: Virtual International Mammalian Synthetic Biology Workshop (mSBW) 2020

Group: Poster Session

Session: General Submission

Authors:

Semeniuk, S., The University of Edinburgh
Liu, W., The University of Edinburgh
Wu, H., The University of Edinburgh
Qian, B., The University of Edinburgh
Cachat, E., University of Edinburgh

Although metastatic cancer is one of the worldâ€™s leading death causes, current knowledge of mechanisms underlying metastasis formation is limited due to the absence of powerful molecular monitoring tools. It is known that macrophages can be reprogrammed by cancer cells to function as anti-inflammatory and pro-tumorigenic cells, giving rise to the multiple subsets of tumour-associated macrophages. One of such subsets is metastasis-associated macrophages (MAMs), which are involved in the tumour establishment at metastatic sites, and, therefore, could act as potential therapeutic targets in metastasis prevention. In this project, we aim at applying a recently developed synthetic Notch receptor (synNotch) to target various macrophage biomarkers to study MAM â€“ cancer cell interactions *in vivo*. This could potentially provide insights into metastasis progression and prevention mechanisms.

Developing a High-Throughput Single-Cell Screening Platform to Optimize Spatial and Temporal Properties of Mammalian Synthetic Circuits and Parts

Conference: Mammalian Synthetic Biology Workshop

Year: 2020

Proceeding: Virtual International Mammalian Synthetic Biology Workshop (mSBW) 2020

Group: Poster Session

Session: General Submission

Authors:

Lee, J., Rice University
Liu, Z., Rice University
Ahrens, J., Rice University
Suzuki, P., Stanford
Lu, X., Rice Unviersity
Lai, S., Baylor College of Medicine
Guan, S., Baylor College of Medicine
St-Pierre, F., Baylor College of Medicine

Biological entities actively move in space and time. Therefore, it is imperative to optimize spatial and temporal properties such as cell-to-cell interaction and gene expression kinetics etc. to successfully repurpose synthetic genetic circuits and parts as therapeutics and diagnostics. Directed evolution is commonly used to screen libraries of variants for desirable properties. However, screening for complex spatial and temporal properties at high-throughput is not available and thereby make the optimization of these properties slow and tedious. To address this problem, we developed SPOTlight, a versatile high-throughput technique to isolate individual mammalian cells with unique spatiotemporal profiles from heterogeneous populations. SPOTlight relies on imaging visual phenotypes by microscopy, precise optical tagging of single target cells, and retrieval of tagged cells by fluorescence-activated cell sorting. We anticipate that the versatility of SPOTlight will facilitate its deployment to rapidly screen for synthetic circuits and parts with optimized spatial and temporal properties.

Design Considerations of a miRNA Sensor

Conference: Mammalian Synthetic Biology Workshop

Year: 2020

Proceeding: Virtual International Mammalian Synthetic Biology Workshop (mSBW) 2020

Group: Poster Session

Session: General Submission

Authors:

Zouein, A., Imperial College London
Ceroni, F., Imperial College London
Bosia, C., Italian Institute for Genomic Medicine
Ellis, T., Imperial College London

Abstract:

MiRNAs play important roles in many diseases such as playing a role in promoting immunosuppressive tumour microenvironments (Ren et al., 2019). As such they have been frequently targeted for biomarker discovery (Condrat et al., 2020). In order to accurately incorporate miRNA response into circuits more information is required to understand mode of action of miRNAs upon internalisation and their effect on native and synthetic mRNA targets. To address these issues, we design a chassis for assembly of synthetic mRNA targets containing varying recognition element (MRE) characteristics. This can then be integrated into a workflow for high throughput iteration of MREs. Therefore, varied sensitivity to miRNAs can be easily tuned. To inform the iterations to be assessed for Cholangiocarcinoma detection, we generate a model that sheds light on favourable MRE characteristics such as spacing, multiplicity and complementarity. The model further assesses sensor sensitivity to varying amounts of miRNA and mRNA target and narrows down the parameters to screen for desired miRNA sensor characteristics.

References:

Ren, W., Zhang, X., Li, W., Feng, Q., Feng, H., Tong, Y., Rong, H., Wang, W., Zhang, D., Zhang, Z., Tu, S., Zhu, X. and Zhang, Q., 2019. Exosomal miRNA-107 induces myeloid-derived suppressor cell expansion in gastric cancer. Cancer Management and Research, Volume 11, pp.4023-4040.

Condrat, C., Thompson, D., Barbu, M., Bugnar, O., Boboc, A., Cretoiu, D., Suciu, N., Cretoiu, S. and Voinea, S., 2020. miRNAs as Biomarkers in Disease: Latest Findings Regarding Their Role in Diagnosis and Prognosis. Cells, 9(2), p.276.

AUTHOR INDEX